关键词

CHINA WATER KEYWORDS

水利部国际经济技术合作交流中心　编著

· 北京 ·

图书在版编目（CIP）数据

中国水利关键词 / 水利部国际经济技术合作交流中心编著. -- 北京 : 中国水利水电出版社, 2025. 3.
ISBN 978-7-5226-3163-9

Ⅰ. TV

中国国家版本馆CIP数据核字第2025SX5829号

书　　名	**中国水利关键词** ZHONGGUO SHUILI GUANJIANCI
作　　者	水利部国际经济技术合作交流中心　编著
出版发行	中国水利水电出版社 (北京市海淀区玉渊潭南路1号D座　100038) 网址：www. waterpub. com. cn E-mail：sales@ mwr. gov. cn 电话：(010) 68545888（营销中心）
经　　售	北京科水图书销售有限公司 电话：(010) 68545874、63202643 全国各地新华书店和相关出版物销售网点
排　　版	中国水利水电出版社微机排版中心
印　　刷	天津嘉恒印务有限公司
规　　格	130mm×184mm　32开本　3.25印张　73千字
版　　次	2025年3月第1版　2025年3月第1次印刷
定　　价	**38.00**元

序言

为贯彻党的二十届三中全会关于“构建更有效力的国际传播体系”的要求，落实李国英部长关于“进一步做好水利双向国际传播工作”的指示，对外讲好中国治水故事，提升水利国际传播效能，推动习近平总书记“节水优先、空间均衡、系统治理、两手发力”治水思路成为国际主流治水理念，《中国水利关键词》以《深入学习贯彻习近平关于治水的重要论述》（中英文版）、《党的二十大报告》（中英文版）、《中华人民共和国国民经济和社会发展第十四个五年规划和 2035 年远景目标纲要》（中英文版）等为基础，系统整理了涉及习近平总书记治水思路、水利高质量发展举措和成效的 370 个水利关键词、283 个水利常用术语及其英文翻译，旨在规范化、标准化水利国际传播的英文翻译用词，为从事水利国际合作和翻译的人员及语言学习者提供参考。

鉴于中国水利高质量发展是一个动态发展过程，因此难免存在疏漏和待完善之处，欢迎广大读者批评指正，我们将在后续修订中逐步完善。

郝钊

2025 年 2 月

目录

序言

水利关键词

水利常用术语

水利关键词

B

1 把水资源、水生态、水环境承载力作为刚性约束

Recognizing the carrying capacity of water resources, water ecology and water environment as rigid constraints

2 把修复长江生态环境摆在压倒性位置

Giving top priority to restoring the eco-environment of the Yangtze River

3 保护传承弘扬黄河文化

Safeguarding, preserving and promoting the Yellow River Culture

4 标志性水利工程

Landmark water projects

5 不伤亡、不漫滩、不垮坝的防御目标

Defensive goal of ensuring no casualties, no floodplain inundation and no dam breaching

6 部门协调联动机制

The mechanism for inter-sectoral coordination and collaboration

C

1. 草原森林河流湖泊休养生息

   Rehabilitation of grasslands, forests, rivers and lakes

2. 产学研用相结合的节水技术创新体系建设

   The establishment of an integrated water-saving technological innovation system based on collaboration between businesses, universities and research institutes

3. 长江防护林体系建设

   The construction of the Yangtze River shelterbelt system

4. 长江干流岸线利用项目清理整治及“回头看”

   Rectification and follow-up inspection of shoreline utilization projects along the trunk stream of the Yangtze River

5. 长江流域岸线保护利用和水资源调配协调机制

   The coordination mechanism for the protection and utilization of the Yangtze River shoreline and for the allocation of the Yangtze River water resources

6. 长江流域联合调度

   The joint operation in the Yangtze River Basin

7 长江流域全覆盖水监控系统

The basin-wide water monitoring system in the Yangtze River Basin

8 长江流域水库群抗旱保供水联合调度专项行动

The specialized program aiming to mitigate the impacts of the drought and ensure water supply by coordinating the reservoir groups in the Yangtze River Basin

9 长江上中游、江河湖库、左右岸、干支流协同治理

Integrated governance of the Yangtze River that covers from source to end, from the left bank to the right bank, both natural and artificial water bodies, and both mainstream and tributaries

10 长江文化

The Yangtze River Culture

11 常态化安全鉴定、除险加固机制

The regular mechanism for safety appraisal, risk removal and structural reinforcement

12 超计划超定额用水

Water use exceeding planned and rationed volume

13 城市防洪能力

Urban flood control capacity

14 城市防洪排涝体系

The urban flood control and drainage system

15 城乡供水安全

Water supply security in urban and rural areas

16 城乡供水和抗旱保障能力

Water supply and drought resistance capabilities in both urban and rural areas

17 城乡供水一体化

The integration of urban and rural water supplies

18 城镇供水管网漏损改造

Upgrades to urban water distribution networks to minimize leakage

19 城镇污水垃圾处理、沿江化工污染治理、农业面源污染治理、船舶污染治理和尾矿库治理专项行动

Special campaigns for rigid control over urban wastewater and solid waste, chemical industrial pollution along rivers, agricultural non-point source pollution and pollution from vessels and tailing ponds

20 初始水权分配制度

The mechanism for initial water rights allocation

21 从观念、意识、措施等各方面都要把节水放在优先位置

Giving priority to water conservation in terms of mindset, awareness, action and others

D

1. 打通国家水网“最后一公里”

   Addressing any remaining gaps in the national water network

2. 大江大河大湖生态保护

   Protecting the eco-environment of major rivers and lakes

3. 大运河国家文化公园数字云平台

   The digital cloud platform of the Grand Canal National Cultural Park

4. 大运河文化

   The Grand Canal Culture

5. 大中型灌区续建配套与现代化改造

   Building support facilities and modernizing existing large and medium-sized irrigation areas

6. 党委领导、政府负责、部门协调、全社会共同参与的水土保持工作格局

   The work pattern for soil and water conservation that features Party committee leadership, governmental responsibility, sectoral synergy and public participation

7 地方人民政府主体责任、水行政主管部门行业监管责任、供水单位运行管理责任“三个责任”体系

The “three-tiered responsibility” system wherein local governments bear the main responsibility for water supply, relevant administrative departments oversee the water supply sector, and water utilities are tasked with the provision and management of the supply

8 地下水超采漏斗区综合治理

Comprehensive management of the funnel-shaped areas affected by excessive groundwater extraction

9 地下水超采治理与保护

Addressing and protecting against groundwater over-extraction

10 地下水利用监测计量体系

Monitoring and measuring system of groundwater use

11 地下水位变化通报机制

The mechanism for notifying changes in groundwater levels

12 地下水自动监测站网

The network of automatic groundwater monitoring stations

13 多水源联合调度

Joint scheduling of multiple water resources

14 多元化水利投融资体制

The diversified water investment and financing system

15 多源联调联供的水资源调配体系

The water resources allocation system that integrates multiple sources

F

1. 防洪减灾能力

   The capacity for flood control and disaster relief

2. 防洪抗旱工程体系

   The flood control and drought relief engineering system

3. 防洪数字化

   Flood control digitalization

4. 防洪智慧化调度

   Intelligent scheduling for flood control

5. 防汛抗旱指挥信息化业务平台

   The flood prevention and drought relief command platform empowered by information technology

6. 防汛抗洪责任制

   The responsibility system for flood prevention and control

7. 非居民用水超定额累进加价制度

   The progressive pricing system for non-residential water use exceeding set quotas

8 废污水处理再生利用

The treatment and recycling of wastewater

9 分区水资源管控体系

Water resources management zoning

10 风暴潮监测预警能力建设

Capacity building for monitoring and early warning of storm surges

G

1. 各方利益协调统一的调度机制体制

Robust institutions and mechanisms that effectively balance and harmonize the interests of all parties involved

2. 根据水资源承载能力优化城市空间布局、产业结构、人口规模

Optimizing urban spatial layout, industrial structure and population size in accordance with the carrying capacity of water resources

3. 工业节水改造

Upgrading industrial water conservation facilities

4. 公共财政水利投入稳定增长机制

The mechanism for steady growth in public financial investment in water sector

5. 共保联动的水生态环境保护与修复体系

The coordinated protection and restoration system for water eco-environment

6 共同抓好黄河大保护，协同推进大治理

Promoting well-coordinated eco-environmental conservation and governance of the Yellow River

7 共抓长江大保护、不搞大开发

Stepping up conservation of the Yangtze River and stopping its excessive development

8 灌区现代化建设与改造

The modernization and transformation of irrigation areas

9 国家、流域、省、市、县五级水旱灾害防御应急响应规程

Emergency response protocols for flood and drought prevention at national, basin, provincial, municipal and county levels

10 国家、省、市三级重点监控用水单位名录

The national, provincial and municipal directories of key water users

11 国家水利风景区

National Water Park

12 国家水情教育基地

National Water Education Base

13 国家防汛抗旱指挥系统

The national flood control and drought relief command system

14 国家骨干网、省级水网、市级水网和县级水网

National backbone water network and provincial, municipal and county-level water networks

15 国家骨干网和省级水网互联互通

The inter-connection and interoperability between the national backbone water network and provincial water networks

16 国家级水权交易平台

The national-level water rights trading platform

17 国家节水行动

The national water conservation campaign

18 国家水安全保障体系

The national water security guarantee system

19 国家水网

The National Water Network

20 国家水网主骨架和大动脉

The framework and main routes of the National Water Network

21 国家水文站网体系

The national hydrological station network

22 国家用水定额体系

The national system for water quotas

H

1. 海绵城市

   Sponge city

2. 海纳百川，有容乃大

   All rivers run into sea, and tolerance brings greatness and respect

3. 旱情监测预警综合平台

   The comprehensive platform for drought monitoring and early warning

4. 合同节水管理

   Contractual water conservation management

5. 河道采砂管理

   River sand mining management

6. 河湖岸线保护与利用规划体系

   The planning system for the protection and utilization of river and lake shorelines

7. 河湖管理保护责任体系

   The responsibility system for river and lake management and protection

8 河湖和湿地生态保护修复

The conservation and restoration of rivers, lakes and wetlands

9 河湖健康生命

The health of rivers and lakes

10 河湖“清四乱”

Cleanups of illegal occupation, mining, dumping and construction for rivers and lakes

11 河湖生态保护治理

The protection and governance of river and lake ecosystems

12 河湖生态补水

Ecological water recharge to rivers and lakes

13 河湖生态流量监测预警

Monitoring and early warning efforts for ecological flows in rivers and lakes

14 河湖生态用水

Ecological water for rivers and lakes

15 河湖湿地萎缩

Shrinkage of rivers, lakes and wetlands

16 河湖水量水质同步监测分析

The synchronous monitoring and analysis of the water quantity and water quality of rivers and lakes

17 河湖水域岸线空间管控

Space management and control over water areas and shorelines of rivers and lakes

18 河湖行泄洪综合治理

The comprehensive management of rivers and lakes with respect to flood passing and release

19 河湖长动态调整和责任递补机制

The dynamic adjustment mechanism of river and lake chiefs with vacancies filled in proper order

20 河湖长制

River and lake chief systems

21 河湖长制考核制度

The performance evaluation framework for river and lake chief systems

22 河湖智慧监管

Smart supervision of rivers and lakes

23 河流伦理

River ethics

24 横向生态保护补偿机制

The inter-provincial ecological protection compensation mechanism

25 洪水保险

Flood insurance

26 洪水资源化利用

Floodwater utilization

27 互联互通的水资源供给保障体系

The interconnected water supply system

28 还水于河

Returning misused ecological water to rivers and lakes

29 黄河大合唱

The Yellow River Cantata

30 黄河干支流防洪体系

The flood control system of the Yellow River's trunk stream and tributaries

31 黄河环境资源巡回法庭

The Yellow River environment and resource circuit court

32 黄河流域生态保护和高质量发展战略

The strategy for the eco-conservation and high-quality development of the Yellow River Basin

33 黄河宁，天下平

Only when the Yellow River is calm, can there be peace in China

34 恢复水清岸绿的水生态体系

Restoring water ecosystem to ensure clear waters and green shores

J

1 极端天气事件

Extreme weather events

2 简政放权、放管结合、优化服务改革

The reform aimed at streamlining administration, delegating powers, integrating decentralization with regulation and optimizing services

3 江河湖泊过度开发

Over-exploitation of rivers and lakes

4 江河控制性工程建设

The building of water control projects on rivers

5 江河战略

The National River Strategy

6 节水贷

Water-saving loans

7 节水护水惜水

Water conservation, protection and cherishing

8 节水关键技术和重大装备

Water-saving technologies and key equipment

9 节水及水循环利用设施

Water-saving and recycling infrastructures

10 节水机制体制改革

Reforms in water conservation systems and mechanisms

11 节水即治污

Water conservation as a means of pollution control

12 节水监督考核工作机制

Water-saving supervision and assessment mechanism

13 节水目标责任制

Water-saving target responsibility system

14 节水先进标杆

Advanced benchmarks for water conservation

15 节约用水工作部际协调机制

The inter-ministerial coordination mechanism for water conservation

16 京杭大运河全线贯通

The full-channel water flow of the Beijing-Hangzhou Grand Canal

17 京杭大运河全线贯通补水行动

The full-channel connectivity and water recharge initiative of the Beijing-Hangzhou Grand Canal

18 精打细算用好水资源，从严从细管好水资源

Maximizing the utilization efficiency of water resources and applying rigorous and meticulous management approaches

K

1. 抗咸潮保供水专项行动

   The special campaign to mitigate the salt tide and ensure the water supply

2. 跨部门综合监管制度

   The cross-departmental comprehensive regulatory collaboration framework

3. 跨地市江河水量分配方案

   Water allocation plans for inter-prefecture rivers

4. 跨流域跨区域重大引调水工程

   Major inter-basin and cross-region water transfer and diversion projects

5. 跨区域管理协调机制

   The mechanism for cross-regional coordination

6. 跨省市河道综合治理

   The comprehensive management of cross-provincial waterways

7 跨行政区河流水系治理保护

The coordination between administrative diversions in river system management and protection

K

L

1 蓝天、碧水、净土保卫战

Making further efforts to keep skies blue, waters clear and lands clean

2 “两个坚持、三个转变”的防灾减灾救灾新理念

The new concept of the “Two Adherences and Three Shifts” for disaster prevention, reduction and relief

3 流域多目标统筹协调调度

The multi-objective operation that covers the entire basin in a concerted and coordinated manner

4 流域管理机构 + 省级河长办协商协作平台

The platform for consultation and cooperation between river basin authorities and river chiefs at the provincial level

5 流域规划实施责任制

The accountability system for the implementation of the basin-wide plans

6 流域内水生态环境保护修复联合防治、联合执法

Joint management and law enforcement for protecting and restoring water environments within river basins

7 流域生命共同体

A community of life in the basin

8 流域生态统一调度

Unified ecological scheduling in the basin

9 流域统筹、区域协同、部门联动的流域治理管理新格局

The framework for effective basin-wide coordination, regional collaboration and cross-departmental interaction at the basin level

10 流域统一规划、统一治理、统一调度、统一管理

Unified planning, governance, scheduling and management of river basins

11 流域协同治理管理机制

The mechanism for synergized governance of river basins

12 流域一体化防洪减灾体系

The integrated flood control and disaster reduction system for river basins

13 流域整体和水资源空间均衡配置

The overall and spatially balanced allocation of water resources in river basins

14 流域治理管理数字化、网络化、智能化

Digitalized, internet-based and smart river basin governance

15 绿色、循环、低碳发展

Green, circular and low-carbon development

16 绿水青山就是金山银山

Lucid waters and lush mountains are invaluable assets

M

1. 民间河长制度

The system of non-governmental river chiefs

2. 民生水利

Putting people's livelihood first in water sector

3. 民生为上、治水为要

Putting people's livelihood first and giving utmost importance to water governance

4. 母亲河复苏行动

The Mother River Revival Action

N

1. 南水北调后续工程高质量发展

   The high-quality development of follow-up projects of the South-to-North Water Diversion Project

2. 年度双控目标完成情况考核监督

   Annual assessment and supervision of the progress of dual regulation targets

3. 农村标准化供水设施建设

   The construction of standard water supply facilities in rural areas

4. 农村供水工程网络

   The rural water supply network

5. 农村供水良性水价形成机制

   A fair pricing mechanism for determining prices for rural water supply

6. 农村集中供水工程

   Centralized rural water supply projects

7 农村集中供水水价改革

The comprehensive reform of centralized rural water supply pricing

8 农村饮水安全

Rural drinking water security

9 农业、工业、服务业国家用水定额

National water quotas for agricultural, industrial and service sectors

10 农业节水增效、工业节水减排、城镇节水降损

Improving agricultural water use efficiency, diminishing industrial water waste and curtailing urban water losses

11 农业农村用水计量体系

The agricultural and rural water metering system

P

1 排污权、用能权、用水权、碳排放权市场化交易

The market-based trading of pollutant discharge rights, energy use rights, water use rights and carbon emission rights

Q

1 区域间基础设施联通

Inter-regional infrastructure connectivity

2 取用水监管机制

The mechanism for water intake and use supervision

3 取水监测计量体系

The water abstraction monitoring and metering system

4 取水口动态更新机制

The dynamic update mechanism for water intakes

5 取水许可及超采限批制度

The system for water abstraction permit and restricted approval of overdraft

6 取用水管理专项整治行动

The special rectification campaign for water intake and use management

7 取用水事中事后监管

Supervision of water use both during and post-abstraction

8 全方位、多层次、立体化监管体系

The all-round, multi-tiered and multi-dimensional regulatory framework

9 全国水利建设市场监管平台

The national water construction market supervision platform

10 全国水利普查

The National Water Census

11 全国水资源调查评价工作

The national water resources survey and evaluation

12 全国用水统计调查直报管理系统

The national water use statistical survey direct reporting management system

13 全过程节水

The whole-process water conservation

14 全球水治理体系

The global water governance system

15 全区域绿色生态水网

The regional green eco-positive water network

16 全生命周期河湖监管

The closed-loop river and lake regulation

17 确有需要、生态安全、可以持续的重大水利工程论证原则

The principle of absolute necessity, ecological security and long-term sustainability of major water projects

R

1. 让河流恢复生命、流域重现生机

   Allowing rivers to restore life and basins to regain vitality

2. 人口经济与资源环境相均衡

   The balance among population growth, economic development, resource utilization and environmental protection

3. 人类命运共同体

   A global community of shared future

4. 人水和谐

   Harmony between humanity and water

5. 人与自然和谐共生

   Harmony between humanity and nature

6. 任职限制、终身禁入和终身追责制度

   Rules regarding restrictions on post, lifetime bans and lifetime accountability

S

1. 三条红线，四项制度（水资源开发利用控制红线、用水效率控制红线、水功能区限制纳污红线；用水总量控制制度、用水效率控制制度、水功能区限制纳污制度、水资源管理责任和考核制度）

The Three Red Lines and Four Systems

The Three Red Lines denote the boundaries for controlling the exploitation and utilization of water resources, the efficiency of water use and the pollution load in water functional areas. The Four Systems encompass the total amount control system for water usage, the water use efficiency control system, the pollution restriction system for water functional areas and the system for water resources management accountability and evaluation.

2. 山洪灾害监测预警平台

The flash flood monitoring and early warning platform

3. 山洪灾害群测群防

Public participation in the monitoring and prevention for flash floods

4 山水林田湖草沙一体化保护和系统治理

Integrated conservation and systematic governance of mountains, waters, forests, farmlands, grasslands and deserts

5 上拦下排、两岸分滞的防洪工程格局

The flood control framework of impounding floods with reservoirs at the upper and middle reaches, regulating floodwater through the use of embankments along river channels and establishing floodplains along both banks of the lower reaches

6 上善若水，水利万物而不争

The greatest virtue is like water, nurturing all living things without competing with them

7 上下游左右岸联防联控机制

The mechanism for collaborative prevention and control between the upper and lower reaches, as well as the left and right banks

8 上蓄、中疏、下排、适当地滞

Storing in the upper stream, dredging in the middle stream, discharging in the lower stream and properly retaining

9 生产建设项目水土保持全链条监督管理体系

The end-to-end supervision and management system for soil and water conservation in production and construction projects

10 生态保护补偿

Compensation for ecological protection

11 生态保护红线

Red lines for ecological protection

12 生态保护修复

Ecological conservation and restoration

13 生态产品价值实现机制

The mechanism for realizing the value of ecological products

14 生态环境保护责任追究制度和环境损害赔偿制度

The accountability system for eco-environmental protection and the system for environmental damage compensation

15 生态清洁小流域

Eco-environmentally clean small watersheds

16 生态系统的完整性和稳定性

Integrity and stability of ecosystems

17 生态系统退化

Ecosystem degradation

18 生态系统服务价值

Value of ecosystem services

19 生态优先，绿色发展

Prioritizing eco-conservation, and pursuing green development

20 省、市、县水网协同融合

The integration of provincial, municipal and county-level water networks

21 省级河湖长联席会议机制

The provincial-level joint meeting mechanism for river and lake chiefs

22 市场化多元化生态补偿机制

The diversified and market-based ecological compensation mechanism

23 收集—发布—督办—问效事中事后监管工作机制

The dynamic "gather-publish-supervise-assess" feedback loop for interim and ex post regulatory oversight

24 数字孪生水利建设

The development of digital twin technology in water sector

25 数字孪生水网

The digital twin water network

26 双圈、两翼、四屏、多廊的水安全保障格局

The water security framework that consists of dual metropolitan areas, two wings, four eco-environmental barriers and multiple eco-environmental corridors

27 水安全保障能力

The overall ability to safeguard water security

28 水风光互补

The complementary between hydropower, wind power and solar power

29 水工程联合调度

The joint operation of water projects

30 水旱灾害防御矩阵

The flood and drought prevention matrix

31 水旱灾害防御能力、水资源节约集约利用能力、水资源优化配置能力、大江大河大湖生态保护治理能力

Capabilities to control and prevent flood and drought disasters, conserve and use water resources intensively and efficiently, optimize the allocation of water resources, and improve the ecological protection and management of major rivers and lakes

32 水旱灾害防御预案体系

The flood and drought prevention planning system

33 水旱灾害应急处置能力

The emergency response capability to deal with droughts and floods

34 水旱灾害预报预警预演预案及调度管理体系

The forecasting, early warning, simulation, preplanning, regulation and management systems for droughts and floods

35 水环境保护

Protecting aquatic environment

36 水环境治理

Water environmental treatment

37 水价水权水市场等重点领域改革

Reforms in pivotal areas such as water pricing, water rights and water markets

38 水价形成机制

The water pricing mechanism

39 水库大坝安全管理责任制

The accountability system for the safety management of reservoirs and dams

40 水库群联合调度

The joint scheduling of the reservoir cluster

41 水利“放管服”改革

The decentralization reform in water sector

42 水利法治体系

The legal system in water sector

43 水利高质量发展

The high quality development in water sector

44 水利工程安全管理责任制

The safety management responsibility system for water projects

45 水利工程标准化管理

Standardized management of water projects

46 水利工程智能化改造

Intelligent renovation of water projects

47 水利规划体系

Water planning system

48 水利基础设施体系

Water infrastructure system

49 水利基础设施投资信托基金

Real estate investment trusts for water infrastructure

50 水利建设市场信用体系建设

The credit system in the water construction market

51 水利精神文明建设

Developing spiritual civilization in water sector

52 水利科技创新能力

Capacity for innovation in water science and technology

53 水利基础设施韧性

Water infrastructure resilience

54 水利是农业的命脉

Water is the lifeline of agriculture

55 水利体制机制法治管理

Institutional frameworks and rule of law in water management

56 水利投融资体制机制

The institutional mechanism for water-related investment and financing

57 水利新发展理念

The new development philosophy in water sector

58 水利新质生产力

The new quality productivity in water sector

59 水利职业院校技能大赛

Skills Competition of Water Vocational Colleges

60 水利遗产保护和认定管理体系

The management system for the protection and recognition of water heritages

61 水利遗产调查

Surveys on water heritages

62 水利治理管理数字化、网络化、智能化

Digital, internet-based and intelligent water governance and management

63 水利综合监管

Integrated water regulation

64 水量调度

Water quantity scheduling

65 水美乡村建设

"Better water for better villages" Program

66 水情信息采集处理、预报预警、防汛抗旱指挥、防洪调度系统

The system for collecting and processing hydrological information, forecasting and early warning, flood prevention and drought relief commanding, and flood control coordination

67 水沙调控调度机制

The mechanism for water scheduling and sediment regulation

68 水生态产品调查监测和价值评价制度

The research, monitoring, and evaluation system for water ecological products

69 水生态产品价值实现机制

The mechanism for realizing the value of water ecological products

70 水生态修复保护

Water ecological protection and restoration

71 水土保持工作协调机制

The coordination mechanism for soil and water conservation

72 水土保持目标责任制和考核奖惩制度

The responsibility system and the performance evaluation framework tied to the goals and targets set for soil and water conservation

73 水土保持数字化

Digitization of soil and water conservation

74 水土保持增汇

Soil and water conservation carbon enhancement

75 水土保持碳汇

Soil and water conservation carbon sequestration

76 水土保持碳汇交易

Soil and water conservation carbon credit trading

77 水土流失和荒漠化石漠化综合治理

The comprehensive control of soil erosion, desertification and stony desertification

78 水网与相关产业协同发展

The coordinated development of water networks and related industries

79 水网智慧化

Intelligent technology application in water network

80 水文化传播

Water culture outreach

81 水文化建设

Water culture development

82 水文化进社区、进机关、进企业、进基层

Water culture activities to communities, government agencies, enterprises and grassroots organizations

83 水污染物肆意排放

Unrestrained pollutants discharge

84 水系连通

Water system connectivity

85 水效标识

Water efficiency labeling

86 水效领跑者行动

Water efficiency leadership initiatives

87 水行政执法

Administrative enforcement in water sector

88 水域面积萎缩

Reduction in water area

89 水源置换

Water source substitution

90 水治理体系和治理能力现代化

The modernization of water governance systems and capabilities

91 水资源保护利用制度

The water conservation and utilization system

92 水资源短缺、水生态损害、水环境污染

Water shortage, water ecological damage and water environmental pollution

93 水资源刚性约束制度

The inelastic water resources constraint system

94 水资源高效利用和合理配置

Efficient use and rational allocation of water resources

95 水资源管理从供给管理向需求管理转变

Shift from water supply management to demand management

96 水资源供给保障体系互联互通

The interconnectivity of the water resource supply guarantee systems

97 水资源供需矛盾

Imbalance between water supply and demand

98 水资源监测计量体系

The water resources monitoring and measurement system

99 水资源监管信息月报机制

The monthly water resources supervision information report mechanism

100 水资源节约保护和优化配置

Conservation, protection and optimal allocation of water resources

101 水资源节约集约利用

The conservation and intensive use of water resources

102 水资源开发利用控制、用水效率控制、水功能区限制纳污红线指标体系

The indicator system for the control of water resources development and utilization, the control of water use efficiency, and pollution limits for water function zones

103 水资源开发利用总体格局

The overall pattern of water resources exploitation and utilization

104 水资源可持续利用

Sustainable use of water resources

105 水资源联合统一调度

Unified scheduling of water resources

106 水资源论证

Water resources assessments

107 水资源配置工程

Water resources allocation projects

108 水资源区域分布不均、年内分配不匀、年际丰枯不定

Uneven regional distribution of water resources, irregular distribution within the year, unpredictable variations in abundance and scarcity from year to year

109 水资源时空分布极不均衡

Extremely uneven spatial and temporal distribution of water resources

110 水资源税改革

Water resources tax reform

111 水资源、水生态、水环境的承载能力

Carrying capacities of water resources, water ecology and water environment

112 水资源统筹调配能力、供水保障能力、战略储备能力

Ability to coordinate the allocation of water resources, guarantee water supply and build strategic water reserves

113 水资源无序利用

Chaotic utilization of water resources

114 水资源有偿使用制度

The system of paid use of water resources

115 水资源总量和强度双控

Dual control of water resources by total quantity and intensity

116 四横三纵、南北调配、东西互济的水资源配置格局

The overall pattern of water resources distribution featuring four rivers in longitude direction (the Yangtze, Yellow, Huaihe and Haihe rivers) and three canals in latitude direction (the eastern, middle and western routes of the South-to-North Water Diversion Project), water diversion from south to north, and mutual supplement between east and west

T

1 滩区居民迁建规划

The resettlement program for residents in the floodplains

2 统筹上下游、左右岸、地上地下、城市乡村

The comprehensive approach that coordinates the upper and lower reaches, left and right banks, surface water and groundwater, and urban and rural areas

3 统筹生产、生活、生态用水

Coordinating the use of water for production, daily life and ecological purposes

4 统筹水灾害、水资源、水环境、水生态治理

Managing water-related disasters, resources, environment and ecology in a coordinated manner

5 统筹水电开发和生态保护

Properly balancing hydropower development and ecological conservation

6 统筹推进灌区骨干工程与高标准农田灌排体系建设

Coordinating the backbone projects in irrigation zones with the establishment of high-standard farmland irrigation and drainage systems

7 统筹做好水灾害防治、水资源节约、水生态保护修复、水环境治理

Coordinating efforts in water disaster prevention and control, water resources conservation, water ecological protection and restoration, and water environmental treatment

8 退耕还林还草

The return of marginal farmlands to forests and grasslands

9 退田还河还湖

Returning farmlands to rivers and lakes

W

1 污水资源化

Sewage as resources

X

1. 习近平总书记“节水优先、空间均衡、系统治理、两手发力”治水思路

   President Xi Jinping's water governance principles of "prioritizing water conservation, balancing spatial distribution, taking systematic approaches and promoting government-market synergy"

2. 系统保护黄河文化遗产

   The systematic preservation of the Yellow River Cultural Heritage

3. 下游河道和滩区环境综合治理

   Cleaning up water courses and floodplains in the lower reaches

4. 夏汛冬枯、北缺南丰

   Summer floods and winter droughts, scarcity in the north and abundance in the south

5. 先节水后调水、先治污后通水、先环保后用水

   Prioritizing water conservation over water diversion, pollution treatment over water supply and environmental protection over water use

6 县、乡、村三级山洪灾害防御预案

Plans for flash flood disaster prevention and control at the county, township and village levels

7 县域节水型社会达标建设

Building a water-saving society at the county level

8 现代生态水网体系

The modern ecological water network system

9 小而美水利项目

Small-scale livelihood water projects

10 小水保安全、大水行蓄洪

Safety in low water and flood conveyance and detention in high water

11 小水电站清理、改造和退出

Rectifying, upgrading and decommissioning of small hydropower stations

12 小水电绿色发展

The green development of small hydropower

13 新建工程水价前置协商机制

The consultation mechanism for water prices before the launch of new construction projects

14 兴利服从防洪、区域服从流域、电调服从水调

Subordinating benefit promotion to flood control, the region to the basin and electricity regulation to water scheduling

15 兴水利、除水害

Harnessing the benefits of water and eliminating its harm

16 幸福河湖

Making more and more rivers and lakes a source of happiness for the people

17 幸福河湖建设

Building rivers and lakes to benefit the people

18 幸福运河

Making the Grand Canal a waterway that benefits the people

19 蓄滞洪区建设和洲滩民垸整治

Constructing flood detention areas and addressing inhabited islets and reclaimed farming polders

20 巡（护）河员

Official river inspectors

Y

1 严守水资源开发利用上限、水环境质量底线和生态保护红线

Strictly adhering to the established limits on water resources exploitation, the baseline for water environmental quality and the red lines for protecting ecosystems

2 一河一策

The “one river, one policy” approach

3 依法治水

Managing water resources in accordance with the law

4 以江河为通道的绿色生态廊道

Eco-environmental corridors with rivers as the main channel

5 以流域为单元，统筹上下游、左右岸、干支流的协调联动机制

The basin-specific coordination and collaboration mechanism coordinating the upper and lower reaches, the left and right banks, and the trunk streams and tributaries

6 以数字孪生流域为核心的智慧水利体系

The smart water system centered on digital twin river basins

7 以水定城、以水定地、以水定人、以水定产

Aligning urban planning, land use, population distribution and production with water availability

8 以水定城、以水定业

Determining urban development and industries based on water availability

9 以水定量、量水而行

Promoting development based on local water conditions, and tailoring measures to local needs

10 以水为师

Learning from the river itself

11 因水制宜、量水而行

Adapting to water conditions and acting according to water availability

12 永定河复苏行动

The Yongding River Revival Initiative

13 永定河贯通入海行动

The Yongding River's Seaward Initiative

14 用水计量在线监测系统

The online water usage monitoring system

15 用水价格形成机制

The pricing mechanism for water use

16 用水统计调查基本单位名录库

The database of water users for water use statistical surveys

17 雨情汛情险情灾情“四情”防御

The defenses against rain, flood, danger and disaster situations

18 雨水情监测预报“三道防线”

The three lines of defense for rainwater monitoring and forecasting

19 预报预警预演预案“四预”措施

The four-sphere preparedness in forecasting, early warning, stimulation and emergency preplanning

Z

1. 战略性水利工程

   Water projects of strategic importance

2. 政府作用和市场机制协同发力

   Government-market coordination

3. 治水安邦、兴水利民

   Water governance to ensure national stability and improve the wellbeing of the people

4. 智者乐水、仁者乐山

   The wise loves water, while the benevolent loves mountains

5. 中华水文化

   Chinese water culture

6. 中国特色水利法治体系建设

   The water governance legal system that reflects China's unique characteristics

7. 中国水故事

   China water stories

8 中华水塔

Water tower of China

9 中小河流治理

The harnessing of small and medium-sized rivers

10 重点跨界水体共保联治

The collective protection and management of critical transboundary water bodies

11 重点流域上下游横向水生态保护补偿机制

The water ecological protection compensation mechanism between upstream and downstream regions in key river basins

12 重点水源、灌区、蓄滞洪区建设和现代化改造

The construction and modernization of key water sources, irrigation areas and flood detention areas

13 珠三角水资源联合调配

Joint allocation of water resources in the Pearl River Delta

14 主要江河防御洪水方案、洪水及水量调度方案、重点水利工程调度运用方案

Comprehensive strategies for flood control of major rivers, scheduling of floodwater and water flows, and scheduling and utilization of key water projects

15 自力更生、艰苦创业、团结协作、无私奉献的红旗渠精神

The indomitable Hongqi Canal Spirit characterized by self-reliance, diligence, solidarity and altruistic commitment

16 用水总量和强度双控、农业节水增效、工业节水减排、城镇节水降损、重点地区节水开源和科技引领创新六大行动

Six major initiatives including dual control of total water consumption and intensity, enhancing efficiency in agricultural water-saving, industrial water-saving and emission reduction, urban water-saving and loss reduction, water-saving and source development in key regions, and leadership by scientific and technological innovation

17 最严格水资源管理制度

The strictest water resources management system

水利常用术语

A

⊙ 安全泄量

Safe discharge volume

B

⊙ 背河地面

Land surface outside embankments

⊙ 崩岸

Bank collapse

⊙ 冰情

Ice jam

⊙ 病险水库

Risky and diseased reservoirs

⊙ 补水

Water recharge

C

⊙ 测雨雷达

Rain radar

⊙ 产水系数

Water yield coefficient

⊙ 潮位

Tide level

⊙ 城市内涝

Urban flooding and waterlogging

⊙ 出库流量

Outflow from reservoirs

D

⊙ 单一水源

Single water source

⊙ 淡水总量

Total freshwater availability

⊙ 堤防险工险段

Critical defense and sections on dikes

⊙ 地表水

Surface water

⊙ 地面沉降

Land subsidence

⊙ 地上悬河

A suspended river

⊙ 地下水

Groundwater

⊙ 地下水超采区

Groundwater overdraft zones

⊙ 地下水回补

Groundwater recharge

⊙ 地下水开采量

Groundwater extraction volume

⊙ 地下水入渗补给

Groundwater recharge by infiltration

⊙ 地下水位

Groundwater level

⊙ 地下水压采

Groundwater overdraft curbing

⊙ 调水

Water diversion

⊙ 调水区

Water origin areas

⊙ 渡槽

Aqueduct

F

⊙ 防洪标准

Flood control standards

⊙ 防洪工程

Flood control project

⊙ 防汛抗旱

Flood control and drought relief

⊙ 非常规水

Non-conventional water

⊙ 非法矮围

Illegal low-rise damming in lakes

⊙ 非法采砂

Illegal sand mining

⊙ 非法取水

Illegal water extraction

⊙ 非法围垦

Illegal reclamation

⊙ 废污水排放

Wastewater discharge

⊙ 分洪工程

Flood diversion project

⊙ 风暴潮

Storm surge

⊙ 富营养化

Eutrophication

G

⊙ 干支流

Mainstream and tributaries

⊙ 高氟水

High-fluorine water

⊙ 供水管网漏损率

Leakage rate of water supply network

⊙ 灌溉排水

Irrigation and drainage

⊙ 灌溉水有效利用系数

Effective utilization coefficient of irrigation water

⊙ 灌排泵站

Irrigation and drainage pumping station

⊙ 灌区

Irrigation districts

⊙ 国际水组织

International water organizations

H

⊙ 海水入侵

Seawater intrusion

⊙ 旱作农业

Rain-fed agriculture

⊙ 涵闸

Culverts and sluices

⊙ 河道断流

Drying-up of river course

⊙ 河道疏浚

River channel dredging

⊙ 河道治理

River training

⊙ 河湖岸线

River and lake bank lines

⊙ 河湖塘沟

Rivers, lakes, ponds and ditches

⊙ 河口

River mouth

⊙ 河势

River flow regime

⊙ 黑臭水体

Black and smelly water bodies

⊙ 洪峰

Flood peak

⊙ 洪水出路

Flood outlet

⊙ 洪水量级

Magnitudes of floods

⊙ 荒漠化

Desertification

J

⊙ 极端干旱

Extreme drought

⊙ 极端强降雨

Extreme rainfall

⊙ 降等报废

Downgrading and decommissioning

⊙ 交叉河道

Intersecting river channels

⊙ 节水灌溉

Water-efficient irrigation

⊙ 节水考核

Water-saving performance evaluation

⊙ 节水科普

Water-saving science popularization

⊙ 节水农业

Water-saving agriculture

⊙ 节水潜力

Water-saving potentials

⊙ 节水认证

Water-saving certification

⊙ 节水型产业

Water-saving industries

⊙ 节水型高校

Water-saving universities

⊙ 节水型企业

Water-saving enterprises

⊙ 节水型社会

Water-saving society

⊙ 节水宣传教育

Promotional and educational campaigns in water conservation

⊙ 节水意识

Water-saving awareness

⊙ 节水载体

water-saving carriers

⊙ 节水指标

Water-saving indicators

⊙ 决口

Breach

K

⊙ 可用水量

Water availability

⊙ 控制断面

Control cross-section

⊙ 口门

River outlet

⊙ 枯季径流

Runoff in dry season

⊙ 苦咸水

Bitter and brackish water

⊙ 库容

Storage capacity of reservoirs

⊙ 垮坝

Dam collapse

⊙ 跨流域

Inter-basin

⊙ 跨区域

Cross-regional

L

⊙ 来沙量

Sediment inflow

⊙ 来水

Inflow

⊙ 蓝藻水华

Cyanobacterial bloom

⊙ 凌汛

Ice-jam flood

⊙ 流域

River basin

⊙ 流域综合规划

Integrated river basin planning

⊙ 漏斗区

Funnel-shaped area

M

⊙ 面源污染

Non-point source pollution

N

⊙ 内河航运

Inland navigation

⊙ 内陆河流

Inland river

⊙ 泥沙

Sediment

⊙ 农村自来水普及率

Tap water penetration rate in rural area

⊙ 农田灌排体系

Farmland irrigation and drainage system

⊙ 农业灌溉水量

Agricultural irrigation water volume

⊙ 农业水价

Agricultural water price

P

⊙ 排涝通道

Flood drainage passageways

⊙ 排水

Water discharge

⊙ 排水沟

Ditch

⊙ 排水系统

Drainage system

⊙ 排污权

Pollutant discharge rights

Q

⊙ 气象干旱

Meteorological drought

⊙ 浅层地下水

Shallow groundwater

⊙ 侵蚀沟

Erosion gully

⊙ 侵占河道

Encroachment on river channels

⊙ 亲水空间

Waterfront spaces

⊙ 秋汛

Autumn flood

⊙ 渠道

Canal

⊙ 取水

Water withdrawal

⊙ 取水口

Water intake

◎ 取用水定额

Water withdrawal and use quotas

R

◎ 人均水资源量

Water resources per capita

◎ 入海口

Estuary

◎ 入河排污口

Sewage outfalls into rivers

S

◎ 山洪

Flash flood

◎ 山洪沟

Flash flood gullies

◎ 墒情

Soil moisture

◎ 上下游

Upstream and downstream

◎ 涉水法律法规

Water laws and regulations

⊙ 深层承压水

Deep confined water

⊙ 深层水回灌

Deep aquifer recharge

⊙ 生态补水

Ecological water recharge

⊙ 生态廊道

Ecological corridor

⊙ 生态流量

Ecological flow

⊙ 生态缺水

Water deficit for ecosystem

⊙ 生态水位

Ecological water level

⊙ 生态修复

Ecological restoration

⊙ 生态堰坝

Eco-friendly barrage

⊙ 生态用水

Ecological water use

⊙ 石漠化

Rocky desertification

◎ 实时水位

Real-time water level

◎ 世界灌溉工程遗产

World Heritage Irrigation Structures

◎ 世界水日

World Water Day

◎ 受水区

Water transfer-in areas

◎ 输配水通道

Water transmission and distribution channels

◎ 输水沿线区

Regions along water transfer route

◎ 水安全

Water security

◎ 水处理

Water treatment

◎ 水法

Water Law

◎ 水费

Water fee

◎ 水害隐患

Water disaster risks

⊙ 水旱灾害

Flood and drought disasters

⊙ 水环境容量

Water environmental capacity

⊙ 水患

Water disasters

⊙ 水价

Water price

⊙ 水库清淤

Dredging of reservoirs

⊙ 水库蓄水量

Reservoir storage capacity

⊙ 水利标准规范

Water standards and specifications

⊙ 水利发电

Hydropower generation

⊙ 水利改革

Reform in water sector

⊙ 水利工程

Water projects

⊙ 水利行业

Water sector

⊙ 水利枢纽

Hydraulic complex

⊙ 水利投融资

Water investment and financing

⊙ 水利遗产

Water heritage

⊙ 水量分配

Water quantity allocation

⊙ 水量分配份额

Water allocation quotas

⊙ 水量损失

Water loss

⊙ 水量消耗

Water consumption

⊙ 水流

Water flow

⊙ 水面萎缩

Water surface shrinkage

⊙ 水情

Water regime

⊙ 水权分配

Water rights allocation

⊙ 水权交易

Water rights trading

⊙ 水沙关系

Water-sediment relationship

⊙ 水生生物多样性

Aquatic biodiversity

⊙ 水生态

Water ecology

⊙ 水生态断面

Water ecology monitoring sections

⊙ 水生态空间

Water ecological space

⊙ 水事活动

Water events and activities

⊙ 水事违法行为

Illegal water activities

⊙ 水体

Water bodies

⊙ 水体流动性

Liquidity of water bodies

⊙ 水体污染

Pollution of water bodies

⊙ 水体自净能力

Self-cleaning capacity of water bodies

⊙ 水挑战

Water challenge

⊙ 水头

Water head

⊙ 水土保持

Soil and water conservation

⊙ 水土流失

Soil erosion

⊙ 水危机

Water crisis

⊙ 水位控制指标

Water level control indicator

⊙ 水位站

Water level station

⊙ 水文

Hydrology

⊙ 水文监测

Hydrological monitoring

⊙ 水文站

Hydrological station

⊙ 水污染

Water pollution

⊙ 水系割裂

Fragmentation of water system

⊙ 水循环

Water cycle

⊙ 水循环利用设施

Water recycling facilities

⊙ 水域

Waters

⊙ 水源保护区

Water source protection area

⊙ 水源工程

Water source project

⊙ 水源涵养区

Water conservation area

⊙ 水闸

Sluices

⊙ 水质达标率

Water quality compliance rate

⊙ 水质断面

Water quality monitoring sections

◎ 水治理

Water governance

◎ 水资源超采地区

Water overdraft areas

◎ 水资源开发利用强度

Intensity of water resources development and utilization

◎ 水资源开发利用总量

Total volume of water resources development and utilization

◎ 水资源利用方式

Water utilization pattern

◎ 水资源利用率

Water utilization ratio

◎ 水资源时空分布

Spatial and temporal distribution of water resources

◎ 水资源税

Water resources tax

T

◎ 台风暴潮

Typhoon storm surge

◎ 滩区

Floodplains

⊙ 碳排放权

Carbon emission rights

⊙ 特大洪水

Extreme floods

⊙ 特种行业用水

Water use in special trades

⊙ 腾退岸线

Vacating shoreline spaces

⊙ 调蓄工程

Regulation and storage projects

⊙ 脱水段

River sections experiencing temporary low flow

W

⊙ 外调水

Transferred-in water

⊙ 万元工业增加值用水量

Water consumption per 10,000 yuan of industrial added value

⊙ 万元国内生产总值用水量

Water consumption per 10,000 yuan of GDP

⊙ 微咸水

Brackish water

⊙ 违规违法取用水

Illegal and illicit water withdrawal or use

⊙ 围堤

Enclosing levee

⊙ 尾矿库治理

Tailings pond management

⊙ 污染物削减

Pollutant reduction

⊙ 污水

Wastewater

⊙ 污水处理设施

Sewage treatment facilities

⊙ 污水排放

Wastewater discharge

⊙ 污水收集率

Wastewater collection rate

⊙ 无证取水

Unlicensed water abstraction

⊙ 物理流域

Physical river basin

X

⊙ 下泄水量

Water discharge volume

⊙ 咸潮上溯

Saltwater intrusion

⊙ 小微水源

Micro-sized to small water source

⊙ 泄排通道

Flood passage and discharge channels

⊙ 泄水

Water discharge

⊙ 新老水问题

Persistent and emerging water issues

⊙ 行洪河道

River channel for flood discharge

⊙ 行洪空间

Flood discharge space

⊙ 蓄洪

Flood retention

⊙ 蓄水

Water storage

⊙ 蓄水池

Water storage pond

⊙ 蓄泄关系

Storage and discharge relationship

⊙ 蓄引提调工程

Water storage, transfer, pumping and diversion projects

⊙ 蓄滞洪能力

Flood retention and detention capacity

⊙ 蓄滞洪区

Flood detention area

⊙ 汛期

Flood season

Y

⊙ 压咸补淡

Replenishing instream freshwater and halting saltwater intrusion

⊙ 遥感监测

Remote sensing monitoring

⊙ 引调水工程

Water transfer and diversion projects

⊙ 饮用水水源地

Drinking water source area

⊙ 应急备用水源

Emergency water source

⊙ 应急管理

Emergency management

⊙ 用水管理

Water use management

⊙ 用水户

Water user

⊙ 用水结构

Water use mix

⊙ 用水强度

Water use intensity

⊙ 用水效率

Water use efficiency

⊙ 用水需求

Water demand

⊙ 用水状况

Water use situation

⊙ 用水总量

Total water consumption

⊙ 淤地坝

Warping dam

⊙ 雨洪资源

Rainwater and floodwater resources

⊙ 雨量站

Rainfall monitoring station

⊙ 雨情

Rainfall dynamics

⊙ 原水水费

Raw water fee

Z

⊙ 灾后救助

Post-disaster rescue and relief

⊙ 灾前预防

Pre-disaster prevention

⊙ 再生水

Recycled water

⊙ 治河历史

History of river governance

⊙ 治水成果

Water governance achievements

⊙ 治水实践

Water governance practices

⊙ 智慧水利

Smart water

⊙ 中水

Reused water

⊙ 洲滩民圩

Inhabited islets and reclaimed farming polders

⊙ 主河槽

Main river channel

⊙ 浊水荒山

Turbid waters and barren hills

⊙ 最小过流能力

Minimum flow capacity

⊙ 左右岸

Left and right banks